USDA

I0038941

United States
Department of
Agriculture

Forest Service

**Northern
Research Station**

General Technical
Report NRS-79

Revised
February 2012

Field Guide to Common Macrofungi in Eastern Forests and Their Ecosystem Functions

**Michael E. Ostry
Neil A. Anderson
Joseph G. O'Brien**

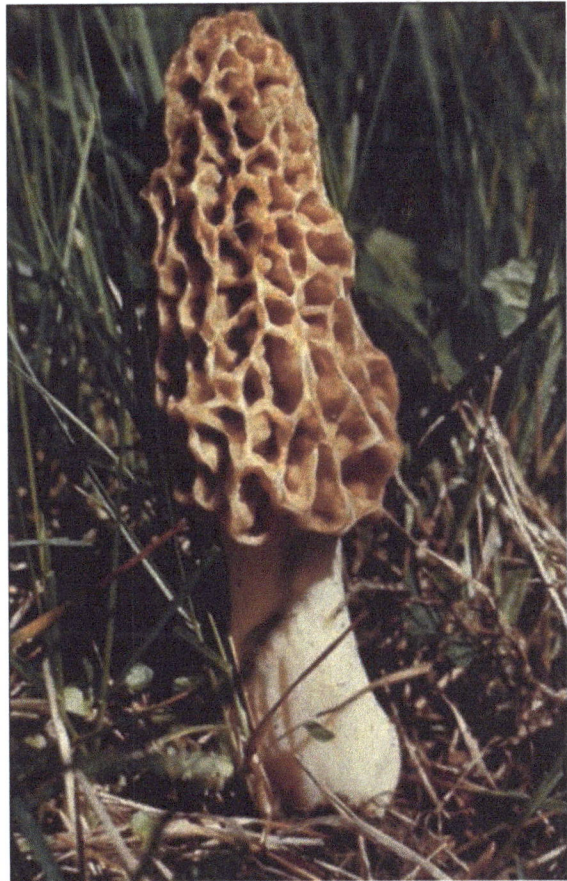

Cover Photos

Front: Morel, *Morchella esculenta*. Photo by Neil A. Anderson, University of Minnesota, used with permission. Back: Bear's Head Tooth, *Hericium coralloides*. Photo by Michael E. Ostry, U.S. Forest Service.

The Authors

MICHAEL E. OSTRY, research plant pathologist, U.S. Forest Service, Northern Research Station, St. Paul, MN

NEIL A. ANDERSON, professor emeritus, University of Minnesota, Department of Plant Pathology, St. Paul, MN

JOSEPH G. O'BRIEN, plant pathologist, U.S. Forest Service, Forest Health Protection, St. Paul, MN

CONTENTS

Upland Conifer Ecosystem

Mycorrhizal
On the ground associated with tree roots

Saprophytic Litter and Wood Decay
On wood

On the ground

INTRODUCTION: ABOUT THIS GUIDE

This guide is intended to serve as a quick reference to selected, common macrofungi (fungi with large fruit bodies such as mushrooms, brackets, or conks) frequently encountered in four broad forest ecosystems in the Midwest and Northeast: aspen-birch, northern hardwoods, lowland conifers, and upland conifers. Although these fungi are most common in the ecosystems we list them in, many can be found associated with tree species in multiple ecosystems. We provide brief identifying characteristics of the selected mushrooms to allow you to identify some down to the species level and others to the genus or group to which they belong. Former scientific names are provided in parentheses. Also included in each mushroom description are details about its ecosystem function, season of fruiting, edibility, and other characteristics.

Although we provide information about edibility in this guide, DO NOT eat any mushroom unless you are absolutely certain of its identity: many mushroom species look alike and some species are highly poisonous. Many mushrooms can be identified only by examining the color of spore prints or by examining spores and tissues under a microscope. As mushrooms age, changes in their shape, color, and general appearance make it necessary to examine several individuals for their distinguishing features.

For additional information on other species of macrofungi, serious mushroom hunters may wish to consult any of the excellent illustrated guides and detailed keys available (see Suggested References at the end of this guide). Several useful mycological Web sites with images and descriptions of fungi are available and a few of these are also listed.

Mushroom Basics

Fungi are important organisms that serve many vital functions in forest ecosystems including decomposition (Fig. 1), nutrient cycling, symbiotic relationships with trees and other plants, biological control of other fungi, and as the causal agents of diseases in plants and animals. Mushrooms are sources of food for wildlife (Figs. 2, 3), and fungi that cause decay in living trees are beneficial to many species of birds and mammals (Figs. 4, 5). Less than 5 percent of the estimated 1.5 million species of fungi have been described, and their exact roles and interactions in ecosystems are largely unknown.

Mike Ostry, U.S. Forest Service

Figure 1.—Mossy Maze Polypore (*Cerrena unicolor* [*Daedalea unicolor*]). Wood decay fungi are critical in nutrient cycling and increasing soil fertility.

Figure 2.—Hollow Stem Larch Suillus (*Suillus cavipes*) stored on branches of black spruce by squirrels.

Figure 3.—Emetic Russula (*Russula emetica*) stored on branches of black spruce by squirrels.

Macrofungi are distinguished from other fungi by their fruiting structures (fruit bodies bearing spores) that we know as mushrooms. Mushrooms with gills, the most common, produce spores that range from white to pink and shades of yellow to brown to black. Most mushrooms produce spores on gills that increase the spore-bearing surface on the underside of the cap. Other mushrooms, such as the Boletes, produce their spores in elongated tubes, and the hedgehog mushrooms produce spores on elongated spines.

Most of the fungus biomass consists of the largely unseen mass of interwoven threadlike hyphae growing in plant tissues and in the soil. Annual variation in the timing and production of the aboveground mushrooms is largely influenced by temperature and precipitation.

The most commonly encountered macrofungi in our woodlands throughout the year are the wood-decaying bracket and conk fungi. These fungi, found on the stems of dead and living trees, produce their spores in small, rigid tubes in leathery-woody fruit bodies that are annual or perennial. The perennial species produce a new layer of tubes to the enlarging fruit body each year.

Figure 4.—Signs of woodpecker activity on aspen decayed by True Tinder Conk (*Fomes fomentarius*) and other decay fungi.

Figure 5.—Cavity in maple decayed by Mossy Maple Polypore (*Oxyporous populinus* [*Fomes connatus*]) used by a squirrel to cache acorns.

Mushroom species form new clones when two compatible spores of the same species germinate and grow together. However, most mushroom spores are dispersed, germinate, and contribute genetic variation to established clones in soil and wood. In nature, many mushrooms and bracket fungi may look alike, but they do not interbreed and thus are distinct biological species. Their growth on different hosts or physical separation from each other over time has made them genetically incompatible.

The parts of a mushroom important for identifying groups and species of fungi are shown in Figure 6. Species of *Amanita* are common, and some are deadly poisonous. Because they possess key identifying parts, we use an *Amanita* to illustrate the key structures of a mushroom.

Young mushrooms are called buttons or the egg stage and contain the preformed cap and stalk. As the mushroom grows, the cap breaks through the egg's universal veil, the stalk elongates and the cap expands

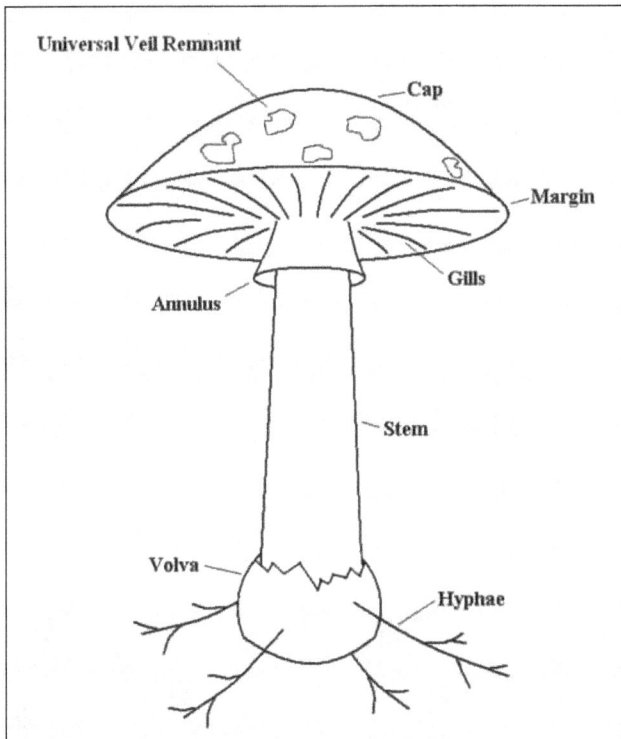

Figure 6.—Identifying parts of an *Amanita* mushroom.
Drawing by Melanie Moore, U.S. Forest Service.

like an umbrella. The secondary veil protecting the gills and spores is broken by the expanding cap, and remnants of this veil form a ring (annulus) on the stem, also referred to as the stipe or stalk. *Amanita* mushrooms also have a cup (volva) at the base of the stalk, often within the soil layer. **Therefore, mushrooms should always be dug, not picked, in order to detect this cup feature of a potentially poisonous mushroom.** In addition to the ring on the stalk and the basal cup, white gills that are free from the stalk and a white spore print distinguish *Amanita* mushrooms from other species.

The fungi illustrated in this guide serve critical ecological functions, and their roles as symbionts, in litter and wood decay, and as pathogens are described. An important beneficial function of many macrofungi is the relationship with forest tree roots in the uptake of nutrients and water and in the protection of the tree roots from pathogenic fungi and nematodes. Strands (hyphae) of the fungus form a dense layer (mantle) around the fine roots of trees and extend out into the surrounding soil. This root-fungus association is called mycorrhizae and benefits both the fungus and the tree.

Pathogenic fungi such as the root and butt rot fungi illustrated in this guide can be damaging, but they also provide important ecological services through nutrient cycling and development of forest structure and wildlife habitat. Distinguishing the potential positive effects from the negative effects of these fungi will enable woodland managers and owners to make informed management decisions based on their objectives.

Mike Ostry, U.S. Forest Service

The honey mushroom (*Armillaria solidipes*) causes a root and butt rot disease of red and white pine, killing trees and creating canopy gaps. However, this increases forest structure and species diversity as trees and woody shrubs resistant to this pathogen regenerate in these gaps.

ASPEN-BIRCH ECOSYSTEM

Aspen-birch

Identification: Cap yellow to orange with white scales that are remnants of the universal veil; white gills free from stalk; white veil; volva (cup) consisting of 2-3 scaly rings on stalk above bulbous base

Season of fruiting: Summer-fall

Ecosystem function: Mycorrhizal with hardwoods and conifers

Edibility: Poisonous

Fungal note: This fungus forms fairy rings that grow radially 3-5 inches every year.

Mike Ostry, U.S. Forest Service

Amanita muscaria

DO NOT eat any mushroom unless you are absolutely certain of its identity.

Destroying Angel

Amanita virosa, A. verna, A. bisporigera

Identification: Cap white, smooth; white gills free from stalk; bulbous base; white veil

Season of fruiting: Summer-fall

Ecosystem function: Mycorrhizal with hardwoods and conifers

Edibility: Highly poisonous and often fatal

Fungal note: These three mushrooms can only be distinguished from each other by their spore characteristics; collectively, they cause 95 percent of fatal mushroom poisonings.

Mike Ostry, U.S. Forest Service

Amanita virosa

DO NOT eat any mushroom unless you are absolutely certain of its identity.

The Omnipresent Laccaria   *Laccaria bicolor*

Identification: Cap colors vary from yellow to buff to orange to lilac, waxy, fibrous; stalk often twisted

Season of fruiting: Summer-fall

Ecosystem function: Mycorrhizal with aspen, spruce, and pine of all ages

Edibility: Good

Fungal note: This is one of the most common mushrooms on upland sites and was the first mycorrhizal fungus to have its entire genome sequenced. *Laccaria longipes* is common with black spruce in bogs.

Neil A. Anderson, University of Minnesota; used with permission

Laccaria bicolor

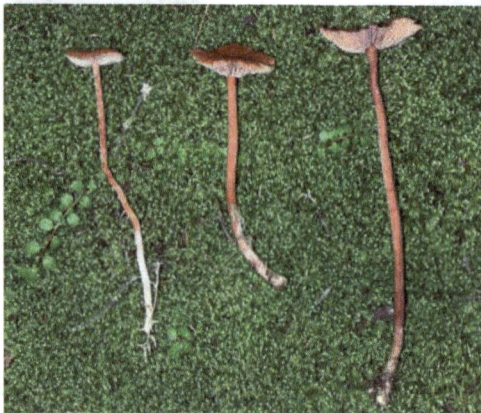

Mike Ostry, U.S. Forest Service

Laccaria longipes

DO NOT eat any mushroom unless you are absolutely certain of its identity.

Aspen Bolete

Leccinum aurantiacum, L. insigne

Identification: Cap red-brown, dry; flesh white turning red then blue-gray when bruised; stalk with brown-black scales called scabers

Season of fruiting: Late summer-fall

Ecosystem function: Mycorrhizal with aspen

Edibility: Edible

Fungal note: Genera of boletes are distinguished from each other by spore color, linear or random arrangement of the tubes on their lower surface, and type of ornamentation on their stalks.

Leccinum aurantiacum

Mike Ostry, U.S. Forest Service

DO NOT eat any mushroom unless you are absolutely certain of its identity.

Birch Bolete   *Leccinum scabrum*

Identification: Cap gray-brown to yellow-brown; flesh white, not staining when bruised; stalk with brown-black scales called scabers

Season of fruiting: Late summer-fall

Ecosystem function: Mycorrhizal with birch

Edibility: Edible

Fungal note: Boletes are among the most sought after edible mushrooms and are ecologically important as tree symbionts.

Joseph O'Brien, U.S. Forest Service

Leccinum scabrum

DO NOT eat any mushroom unless you are absolutely certain of its identity.

Oyster Mushroom

Pleurotus populinus (*P. ostreatus*)

Identification: Cap white-pale tan; stem usually lateral or absent; gills white and run down the stem; spore print white; found only on aspen

Season of fruiting: Summer-fall

Ecosystem function: Sapwood rotter

Edibility: Choice

Fungal note: Three closely related species are known: *P. populinus* is found on aspen; *P. pulmonarius* (*P. sapidus*) is found on hardwoods other than aspen and has a lilac color spore print; *P. ostreatus* has a white spore print and is also found on hardwoods other than aspen, often in riparian areas.

Mike Ostry, U.S. Forest Service

Pleurotus populinus

DO NOT eat any mushroom unless you are absolutely certain of its identity.

Artist's Conk

Ganoderma applanatum
(Fomes applanatus)

Identification: Shelf-like, hard gray-brown zonate upper surface; white lower surface that turns brown when scratched

Season of fruiting: Perennial

Ecosystem function: Causes a white stem and butt rot of hardwoods

Edibility: Inedible

Fungal note: The most common perennial wood decay fungus of dead and dying hardwood trees. A single conk can produce 1.25 billion spores each hour for 5-6 months each year.

Mike Ostry, U.S. Forest Service

Ganoderma applanatum

False Tinder Conk

Phellinus tremulae (*Fomes ignarius*)

Identification: Hoof-shaped, gray-black hard conk with a brown margin

Season of fruiting: Perennial

Ecosystem function: Causes a white trunk rot of aspen

Edibility: Inedible

Fungal note: This fungus causes more wood volume loss than any other aspen pathogen; however, the resulting soft wood of affected stems is beneficial for cavity-nesting wildlife. On average, decay extends 8 feet above and 5 feet below an individual conk.

Mike Ostry, U.S. Forest Service

Phellinus tremulae

DO NOT eat any mushroom unless you are absolutely certain of its identity.

Cross section of aspen stem near a conk of False Tinder Conk (*P. tremulae*) revealing a column of soft, decayed wood that benefits cavity-nesting birds and animals.

Cavities excavated by woodpeckers in aspen affected by *P. tremulae*.

DO NOT eat any mushroom unless you are absolutely certain of its identity.

True Tinder Conk   *Fomes fomentarius*

Identification: Hoof-shaped, gray, hard conk

Season of fruiting: Perennial

Ecosystem function: Causes a wood rot, common on dead birch

Edibility: Inedible

Fungal note: The felt-like inner layer makes excellent tinder. This material, called "amadou," has also been used as a substitute for matches after soaking it in solutions of potassium or sodium nitrate and then drying it.

Mike Ostry, U.S. Forest Service

Fomes fomentarius

DO NOT eat any mushroom unless you are absolutely certain of its identity.

Birch Polypore

Piptoporus betulinus
(*Polyporus betulinus*)

Identification: Circular, soft to leathery, shelf-like, white to brown

Season of fruiting: Annual

Ecosystem function: Causes a brown cubical wood rot, common on dead birch trees

Edibility: Tough, inedible unless very young

Fungal note: The inner material of the conk can be used as fire tinder when dry.

Piptoporus betulinus

DO NOT eat any mushroom unless you are absolutely certain of its identity.

Multicolor Gill Polypore — *Lenzites betulina*

Identification: Fruit body leathery, hairy with alternating bands of gray, yellow, and brown; undulating, gray gills

Season of fruiting: Summer-fall

Ecosystem function: White sapwood rot of dead birch and other hardwoods

Edibility: Inedible

Fungal note: Fruit bodies are white when young, turning gray with age, often with green algae on the surface.

Lenzites betulina, bottom view.

Morel (Sponge Mushroom) *Morchella esculenta*

Identification: Cap resembles an inverted pine cone with ridges and deep pits, gray-cream-yellow; stem white-cream and hollow

Season of fruiting: Brief (2-3 weeks) in spring

Ecosystem function: Litter and wood decay; found on the ground among aspen and many other hardwood species, spruce, and pine of all ages

Edibility: Choice

Fungal note: One of the most sought after edible mushrooms.

Morchella esculenta

DO NOT eat any mushroom unless you are absolutely certain of its identity.

Honey Mushroom   *Armillaria gallica*

Identification: Cap tan to golden yellow; prominent ring on stem; white spore print; black "shoestring" cords (rhizomorphs) that transport food to growing hyphae

Season of fruiting: Fall

Ecosystem function: Root and butt rot capable of killing trees, especially stressed trees, creating root rot pockets resulting in canopy gaps

Edibility: Choice

Fungal note: An individual clone of this fungus, 15.4 ha in size and estimated to be 1,500 years old, was identified in northern Michigan. The mushroom *Entoloma abortivum* parasitizes fruit bodies of *Armillaria* turning them into misshapen Abortive *Entoloma* mushrooms.

Joseph O'Brien, U.S. Forest Service

Armillaria gallica

DO NOT eat any mushroom unless you are absolutely certain of its identity.

Armillaria sp. with characteristic ring (annulus) on the stems.

Abortive *Entoloma* fruit body resulting from *Armillaria* mushrooms parasitized by *Entoloma abortivum*.

DO NOT eat any mushroom unless you are absolutely certain of its identity.

NORTHERN HARDWOOD ECOSYSTEM

Northern hardwoods

Giant Puffball   *Calvatia gigantea*

Identification: Softball-soccer ball in size; white leathery skin when young turning yellow-tan when mature

Season of fruiting: Late summer-fall

Ecosystem function: Mycorrhizal

Edibility: Edible when young

Fungal note: Giant puffballs 30.5 cm in diameter can produce 7 trillion or more spores that are perfectly adapted to wind dissemination. In calm air, spores fall at a rate of 0.5 mm per second.

Joseph O'Brien, U.S. Forest Service

Calvatia gigantea

DO NOT eat any mushroom unless you are absolutely certain of its identity.

Calvatia gigantea

DO NOT eat any mushroom unless you are absolutely certain of its identity.

Bear's Head Tooth   *Hericium coralloides*

Identification: From a single stem, the fruit body branches into clusters of snow-white spines that point down and bear the spores of the fungus on their outer surface. Spines darken to yellow or brown with age.

Season of fruiting: Late summer-fall

Ecosystem function: Decay of hardwood logs

Edibility: Choice

Fungal note: This fungus can be pickled, marinated, or fried.

Joseph O'Brien, U.S. Forest Service

Hericium coralloides

DO NOT eat any mushroom unless you are absolutely certain of its identity.

Scaly Pholiota   *Pholiota squarrosa*

Identification: Cap dry, yellow-pale tan with brown scales; gills yellow-light brown, brown spores; stalk with a veil forming a ring, scales present below but not above ring

Season of fruiting: Summer-fall

Ecosystem function: Wood rotter of hardwoods and conifers

Edibility: Not recommended

Fungal note: Common butt rotter of living aspen and birch as well as down aspen logs. Often found in large clusters. *P. squarrosoides* is another very similar *Pholiota* species that is frequently found.

Mike Ostry, U.S. Forest Service

Pholiota squarrosa

DO NOT eat any mushroom unless you are absolutely certain of its identity.

Milk-White Toothed Polypore *Irpex lacteus* (*Polyporus tulipiferae*)

Identification: White, crust-like, flat to substrate, pores breaking into teeth

Season of fruiting: Spring-fall

Ecosystem function: White rot of hardwoods

Edibility: Inedible

Fungal note: This fungus is very common on dead branches of hardwood trees.

Irpex lacteus This specimen has discolored to yellow-brown with age.

Violet Polypore
Trichaptum biforme
(*Hirschioporus pargamenus, Polyporus pargamenus*)

Identification: Fruit bodies thin, leathery, with zones of various colors and a violet pore surface only on the fruit body margin that breaks into teeth with age; often covering large areas of dead trees.

Season of fruiting: Spring-fall

Ecosystem function: White pocket rot of hardwoods, very common on dead aspen; a very similar species, *T. abietinum*, occurs on conifers

Edibility: Inedible

Fungal note: One of the most common decay fungi in the U.S. The sporocarps are often covered with green algal growth.

Trichaptum biforme

Joseph O'Brien, U.S. Forest Service

DO NOT eat any mushroom unless you are absolutely certain of its identity.

Trichaptum biforme, upper and lower surface

Early growth form of Violet Polypore (*Trichaptum biforme*) on the lower surface of a fallen aspen stem.

DO NOT eat any mushroom unless you are absolutely certain of its identity.

Smoky Polypore

Bjerkandera adusta (*Polyporus adustus*)

Identification: Clusters, small, white to grayish, velvety caps, pore surface gray to black

Season of fruiting: Spring-fall

Ecosystem function: White sapwood rot of dead hardwood trees

Edibility: Inedible

Fungal note: Fruit bodies can revive after long periods of drought. Pores of fruit body are very small (5-7 pores per millimeter).

Mike Ostry, U.S. Forest Service

Bjerkandera adusta

Bjerkandera adusta

Bjerkandera adusta

DO NOT eat any mushroom unless you are absolutely certain of its identity.

Common Split Gill   *Schizophyllum commune*

Identification: Clusters of leathery, whitish gray, fan-shaped gilled fruit bodies

Season of fruiting: Perennial

Ecosystem function: White sapwood rot of living and dead hardwood trees

Edibility: Inedible

Fungal note: Spores of this fungus were obtained from fruit bodies after 50 years of dry storage. Each gill is split into two halves that curl in dry weather to protect the spore-bearing surface.

Mike Ostry, U.S. Forest Service

Schizophyllum commune

Mike Ostry, U.S. Forest Service

Schizophyllum commune, lower gill surface.

DO NOT eat any mushroom unless you are absolutely certain of its identity.

Hen of the Woods

Grifola frondosa
(Polyporus frondosus)

Identification: Large, dull white to gray, solitary fruit bodies with overlapping shelves on the ground near stumps or at the base of living hardwood trees

Season of fruiting: Late summer-fall

Ecosystem function: White butt rot of hardwoods

Edibility: Choice

Fungal note: Always found growing on the ground.

Grifola frondosa (top view)

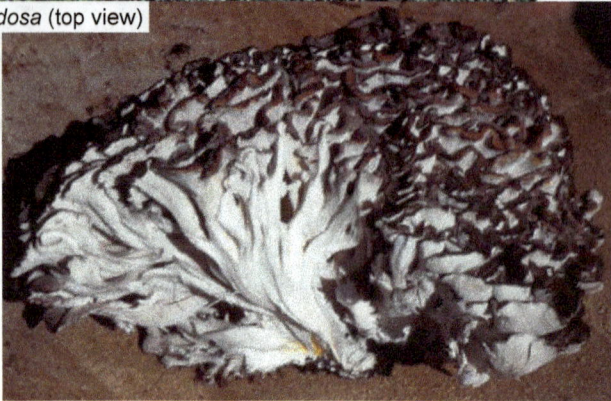

Grifola frondosa (bottom view)

DO NOT eat any mushroom unless you are absolutely certain of its identity.

Maze Bracket   *Daedalea quercina*

Identification: Gray to light brown, leathery, shelf with mazelike lower surface

Season of fruiting: Spring-fall

Ecosystem function: Brown heart rot of oaks

Edibility: Inedible

Fungal note: This fungus is not found west of the Mississippi River.

Joseph O'Brien, U.S. Forest Service

Daedalea quercina, bottom view.

DO NOT eat any mushroom unless you are absolutely certain of its identity.

Annual Shelf Fungus

Phellinus gilvus
(*Polyporus gilvus*)

Identification: Leathery, yellow to brown shelf, yellow-brown interior

Season of fruiting: Summer-fall

Ecosystem function: White sapwood decay and occasionally heart rot

Edibility: Inedible

Fungal note: Common on red oak and other hardwood trees.

Neil A. Anderson, University of Minnesota; used with permission

Phellinus gilvus

Hoof Conk

Phellinus everhartii
(*Fomes everhartii*)

Identification: Woody, hoof-shaped, brown to black and crusty upper surface, rusty brown interior

Season of fruiting: Perennial

Ecosystem function: White heart rot

Edibility: Inedible

Fungal note: Common on oaks, this fungus can cause large economic losses.

Phellinus everhartii

Diamond Polypore

Polyporus alveolaris
(Favolus alveolaris)

Identification: Fruit body cream to orange or reddish brown; short lateral stalk, white to buff color; large diamond-shaped tubes

Season of fruiting: Spring-early summer on dead hardwood branches

Ecosystem function: White rot

Polyporus alveolaris, bottom view.

Edibility: Edible when young

Fungal note: Can cause decay when wood is at low moisture content.

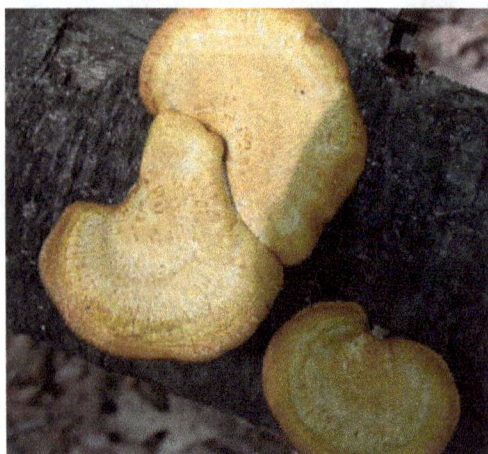

Polyporus alveolaris, top view.

DO NOT eat any mushroom unless you are absolutely certain of its identity.

Dryad's Saddle   *Polyporus squamosus*

Identification: Fan-shaped with central stalk, white-yellow with brown scales, white pore surface

Season of fruiting: Spring-fall

Ecosystem function: White heart rot of hardwood trees

Edibility: Inedible

Fungal note: It was estimated that a single fruit body of this fungus could produce 100 billion spores.

Polyporus squamosus

DO NOT eat any mushroom unless you are absolutely certain of its identity.

Wildlife cavity in elm with heart rot caused by Dryad's
Saddle (*Polyporus squamosus*).

DO NOT eat any mushroom unless you are absolutely certain of its identity.

Cinnabar-Red Polypore *Pycnoporus cinnabarinus (Polyporus cinnabarinus)*

Identification: Orange-red, broadly attached leathery cap

Season of fruiting: Summer-fall

Ecosystem function: White sapwood rot of dead hardwoods

Edibility: Inedible

Fungal note: Some fruit bodies can produce spores into the second and third years.

Joseph O'Brien, U.S. Forest Service

Polyporus cinnabarinus, top view.

Joseph O'Brien, U.S. Forest Service

Polyporus cinnabarinus, bottom view.

DO NOT eat any mushroom unless you are absolutely certain of its identity.

Turkey Tail

Trametes versicolor
(*Coriolus versicolor,*
Polyporus versicolor)

Identification: Cap thin, leathery bracket-like; surface velvet-like with concentric bands of brown-red-yellow-gray-blue colors; pores white-yellow

Season of fruiting: Spring-fall

Ecosystem function: Causes a white rot of hardwood trees and logs

Edibility: Inedible

Fungal note: Wood decayed by this fungus often has black zone lines where different clones of this species meet but do not exchange genetic material. The zone lines produce beautiful patterns in turned vases and other objects made with the affected wood, known as spalted wood.

Mike Ostry, U.S. Forest Service

Trametes versicolor

DO NOT eat any mushroom unless you are absolutely certain of its identity.

Weeping Polypore — *Ischnoderma resinosum (Polyporus resinosus)*

Identification: Clusters of shelf-like fruit bodies; surface dark brown and velvety with a broad white margin; amber drops of a watery fluid on the surface when fresh

Season of fruiting: Summer-fall

Ecosystem function: Decay of hardwoods, causes a white rot of sapwood and heartwood that causes the annual rings to separate

Edibility: Inedible

Fungal note: The pores of older fruit bodies break up into tooth-like spines and the entire fruit body becomes brittle. The fruit body has an anise-like odor. A very similar form of this species occurs on conifers.

Joseph O'Brien, U.S. Forest Service

Ischnoderma resinosum

DO NOT eat any mushroom unless you are absolutely certain of its identity.

Coral-Like Jelly Fungus

Tremellodendron pallidum

Identification: Fruit body resembling coral with white, leathery, flattened upright branches found on the ground in hardwood and conifer stands

Season of fruiting: Summer-fall

Ecosystem function: Decay of litter

Edibility: Inedible

Fungal note: The spores of the true coral fungi develop on structures (basidia) on the exterior of their branches while spores of *Tremellodendron pallidum* develop on basidia within the branches.

Mike Ostry, U.S. Forest Service

Tremellodendron pallidum

DO NOT eat any mushroom unless you are absolutely certain of its identity.

Northern Tooth

Climacodon septentrionalis (Steccherinum septentrionale)

Identification: Overlapping yellowish-white annual shelves with toothed undersides found on living hardwoods, especially maples

Season of fruiting: Late summer-fall

Ecosystem function: Spongy heart rot

Edibility: Inedible

Fungal note: This fungus fruits only occasionally on individual trees, and its teeth can reach 10-15 mm in length.

Joseph O'Brien, U.S. Forest Service

Climacodon septentrionalis

DO NOT eat any mushroom unless you are absolutely certain of its identity.

Inky Caps

Coprinus, Coprinellus, Coprinopsis spp.

Identification: Cap conical in shape, tissue autodigests from the gills and cap margin into a black liquid containing black spores. *Coprinus comatus* (shaggy mane) has a large, white, scaly columnar cap; *Coprinellus micaceus* (mica cap) has a brown cap with mica-like particles; *Coprinopsis atramentaria* has a light gray-brown cap and occurs in clusters of 3 or more.

Season of fruiting: Summer-fall

Ecosystem function: This group of fungi fruits on buried, decayed woody debris

Edibility: Edible

Fungal note: *Coprinopsis atramentaria* and probably other related species contain coprine, a toxin that interacts with alcohol when ingested and causes severe nausea.

Joseph O'Brien, U.S. Forest Service

Coprinus comatus

DO NOT eat any mushroom unless you are absolutely certain of its identity.

UPLAND CONIFER ECOSYSTEM

Mike Ostry, U.S. Forest Service

Upland conifer

Slippery Jack Bolete   *Suillus luteus*

Identification: Cap smooth, sticky red-brown; flesh white; tube openings radiate out from stalk in a linear pattern

Season of fruiting: Late summer-fall

Ecosystem function: Mycorrhizal with red pine of all ages

Edibility: Edible after removing skin of cap

Fungal note: Boletes are important food for insect larvae, invertebrates, turtles, snails, slugs, and many mammals, especially squirrels who often store the mushrooms in trees. Another bolete, *S. brevipes*, often found with jack pine, has such a short stalk that it looks like the cap is resting directly on the ground.

Suillus luteus, top and bottom views.

White Pine Bolete   *Suillus americanus*

Identification: Cap yellow with red streaks, smooth; flesh yellow; tube openings radiate out from stalk in a linear pattern

Season of fruiting: Late summer-fall

Ecosystem function: Mycorrhizal only with white pine

Edibility: Edible

Fungal note: In mixed plantings of red and white pine, this mushroom will be found only in association with white pine. *S. luteus* will be found fruiting under red pine usually at the same time or within 1-2 weeks of *S. americanus*.

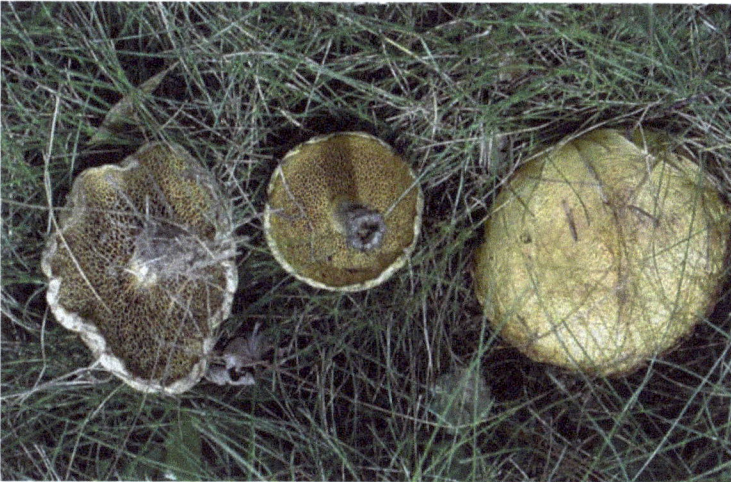

Mike Ostry, U.S. Forest Service

Suillus americanus, top and bottom views.

DO NOT eat any mushroom unless you are absolutely certain of its identity.

The King Bolete   *Boletus edulis*

Identification: Cap cream-brown to reddish brown; tube openings random resembling a sponge; flesh white-yellow; stalk white-ivory with fine lines (reticulations) forming a net

Season of fruiting: Late summer-fall

Ecosystem function: Mycorrhizal with pine, spruce, oak, and birch

Edibility: Choice

Fungal note: Although this bolete is well-known in Europe, research suggests that there are different strains of this species varying in color and shape in North America.

Boletus edulis

DO NOT eat any mushroom unless you are absolutely certain of its identity.

Boletus edulis

Boletus edulis

DO NOT eat any mushroom unless you are absolutely certain of its identity.

Golden Chanterelle   *Cantharellus cibarius*

Identification: Cap yellow to orange, funnel-shape; gills shallow, yellow, blunt, and run down the yellow stalk

Season of fruiting: Summer-fall

Ecosystem function: Mycorrhizal with pine and upland hardwoods

Edibility: Choice

Fungal note: The *Cantharellus* mushrooms are known worldwide as chanterelles and are some of the very best edible mushrooms. Chanterelles are always found growing from soil, unlike false chanterelles (*Hygrophoropsis aurantiaca*) that are found on woody debris.

Joseph O'Brien, U.S. Forest Service

Cantharellus cibarius

DO NOT eat any mushroom unless you are absolutely certain of its identity.

Identification: Cap light lilac in color; annulus is curtain-like (individual threads are distinct)

Season of fruiting: Late summer-fall

Ecosystem function: Mycorrhizal with conifers

Edibility: Highly poisonous. No species in this genus should be eaten because some contain a deadly toxin.

Fungal note: The curtain-like annulus covering the gill surface is a distinctive trait of this genus.

Neil A. Anderson, University of Minnesota; used with permission

Cortinarius traganeus

Mike Ostry, U.S. Forest Service

Curtain on young *Cortinarius.*

DO NOT eat any mushroom unless you are absolutely certain of its identity.

Identification: Mature fruit body the color and shape of a small russet potato with a chambered interior, white when young, form below or at the soil surface

Season of fruiting: Fall

Ecosystem function: One of the most important mycorrhizal species with red pine of all ages

Edibility: Inedible

Fungal note: More than 200 species of *Rhizopogon* have been described. They are eaten and inadvertently spread by many wildlife species.

Mike Ostry, U.S. Forest Service

Rhizopogon sp.

False Chanterelle   *Hygrophoropsis aurantiaca*

Identification: Cap orange to orangish-brown, shallow, velvety, funnel-shaped; gills attached to stem; flesh waxy

Season of fruiting: Late summer-fall

Ecosystem function: Decay of woody debris

Edibility: Not recommended

Fungal note: Often mistaken for the true chanterelle (*Cantharellus cibarius*), but the true chanterelle is a soil fungus and does not grow on woody debris.

Hygrophoropsis aurantiaca, top and bottom views.

Mike Ostry, U.S. Forest Service

DO NOT eat any mushroom unless you are absolutely certain of its identity.

Pine Conk   *Phellinus pini (Fomes pini)*

Identification: Shelf-like, tough, red-brown to brown-black

Season of fruiting: Perennial

Ecosystem function: Causes white pocket rot of living pine

Edibility: Inedible

Fungal note: This fungus causes more decay of living pines than any other fungus but does not decay wood in service such as poles, posts, and structural timbers.

Phellinus pini

Joseph O'Brien, U.S. Forest Service

The Red Band Fungus   *Fomitopsis pinicola*
                                   (Fomes pinicola)

Identification: Brown-black, crusty fruit body with white-red margin and yellow-brown lower pore surface

Season of fruiting: Perennial on conifers and hardwoods

Ecosystem function: Common on dead trees and logs causing a brown rot

Edibility: Inedible

Fungal note: Several biological species of this fungus are known.

Mike Ostry, U.S. Forest Service

Fomitopsis pinicola

DO NOT eat any mushroom unless you are absolutely certain of its identity.

Conifer Parchment

Phlebiopsis gigantea
(Peniophora gigantea)

Identification: Thin, white-tan crust on stumps and logs of pine that still have bark on them

Season of fruiting: Perennial

Ecosystem function: Early colonizer of conifer sapwood

Edibility: Inedible

Fungal note: This is the world's best known biological control fungus. Conidia naturally disseminated or purposely applied onto freshly cut pine stumps will prevent decay by *Heterobasidon annosum*. Widely used in Europe, its use is not yet approved in the U.S.

Mike Ostry, U.S. Forest Service

Phlebiopsis gigantea on pine stump.

Velvet Top Fungus

Phaeolus schweinitzii (*Polyporus schweinitzii*)

Identification: Cap is a shallow funnel with a central stalk when decaying roots or in the form of a bracket when decaying standing trees, stumps, and logs. Color ranges from yellow-brown to dark red-brown, hairy with concentric ridges. Pores form a maze when young, becoming toothed with age.

Season of fruiting: Summer-fall

Ecosystem function: Decays the heartwood of living and dead red pine

Edibility: Inedible

Fungal note: Mature red pines affected by this fungus are commonly wind thrown.

Mike Ostry, U.S. Forest Service

Phaeolus schweinitzii

DO NOT eat any mushroom unless you are absolutely certain of its identity.

False Morel — *Gyromitra esculenta*

Identification: Cap red-brown, irregular, brain-like; stalk white-yellow

Season of fruiting: Spring

Ecosystem function: Litter fungus in red and jack pine stands

Edibility: Poisonous; fumes while boiling this fungus can be toxic

Fungal note: This fungus is reported to produce the compound mono methyl hydrazine, found in rocket fuel.

Mike Ostry, U.S. Forest Service

Gyromitra esculenta

DO NOT eat any mushroom unless you are absolutely certain of its identity.

Witches Hat

Hygrocybe conica
(Hygrophorous conicus)

Identification: Cap cone-shaped with a definite peak when young, golden yellow-orange or red, sticky when wet; gills are waxy, white to olive yellow, and almost free from the stalk that is often twisted, hollow, striated and the same color as the cap

Season of fruiting: Summer-fall

Ecosystem function: Decays litter in conifer and hardwood stands. Some *Hygrocybe* species may be mycorrhizal.

Edibility: Not recommended

Fungal note: These brilliantly colored mushrooms have waxy gills that are triangular in cross section.

Neil A. Anderson, University of Minnesota, used with permission

Hygrocybe conica

DO NOT eat any mushroom unless you are absolutely certain of its identity.

Identification: Club-shaped, smooth, orange-dull yellow

Season of fruiting: Summer-fall

Ecosystem function: Decay of pine litter

Edibility: Inedible

Fungal note: A similar fungus, *C. pistillaris*, is found decaying litter in hardwood stands.

Clavariadelphus ligula

Honey Mushroom

Armillaria solidipes
(A. ostoyae)

Identification: Cap golden yellow; prominent ring on stem; black shoe-string cords (rhizomorphs) under bark of infected trees or in the soil

Season of fruiting: Late summer-fall

Ecosystem function: Causes a root and butt rot of pine

Edibility: Edible

Fungal note: The genus *Armillaria* is complex and contains 10 biological species that have restricted geographical distributions and vegetation associations. Species can be distinguished only by using laboratory techniques. All species are luminescent, often glowing in patches of decayed root or stem tissue. Clones of *Armillaria* several hundred acres in size have been found in the western U.S.

Mike Ostry, U.S. Forest Service

Armillaria solidipes

DO NOT eat any mushroom unless you are absolutely certain of its identity.

Truffle Eater   *Cordyceps ophioglossoides*

Identification: Club-shaped; yellow to olive-brown; yellow threads extending down into the soil where it parasitizes the fungus *Elaphomyces granulatus* (deer truffle)

Season of fruiting: Fall

Ecosystem function: Parasite of the deer truffle (a dark brown sphere with thick walls and a solid black interior found underground) that is mycorrhizal with jack and red pine. Other *Cordyceps* species are parasites of insect larvae and aboveground plant feeding aphids.

Edibility: Inedible

Fungal note: From a 1-m^2 sample area in a Minnesota jack pine stand, it was estimated that there were about 410,000 deer truffles per ha. These truffles are fed upon by many mammal species.

Mike Ostry, U.S. Forest Service

Cordyceps ophioglossoides the truffle eater. Several deer truffles (Elaphomyces granulatus) were dug nearby and placed in the foreground.

DO NOT eat any mushroom unless you are absolutely certain of its identity.

Conifer-Base Polypore *Heterobasidion irregulare*
(*H. annosum, Fomes annosus*)

Identification: Fruit body small, white, "popcorn-like," later lying flat (resupinate) or shelf-like at base of trunks or on stumps; upper surface dark brown to black, hairy, becoming smooth with a hard crust; pore surface white-yellow

Season of fruiting: Perennial

Ecosystem function: Causes a root and butt rot

Edibility: Inedible

Fungal note: *H. annosum* is a species complex with pine, spruce, or fir hosts. *Phlebiopsis gigantea* (conifer parchment) is used as a natural biological control of *H. annosum* in Europe when commercial formulations are applied to fresh stumps of pine when stands are thinned.

Mike Ostry, U.S. Forest Service

Heterobasidion irregulare at base of tree.

DO NOT eat any mushroom unless you are absolutely certain of its identity.

Heterobasidion irregulare, pore surface.

Heterobasidion irregulare, on young white pine.

DO NOT eat any mushroom unless you are absolutely certain of its identity.

Sulfur Shelf

Laetiporus sulphureus (Polyporus sulphureus)

Identification: Multiple clusters of yellow-orange shelves growing on wood, soft, fleshy when young, turning hard when mature

Season of fruiting: Summer-fall

Ecosystem function: Causes a brown cubical rot of living and dead hardwood and conifer trees

Edibility: Edible when young

Fungal note: This fungus, also called chicken of the woods, is very common on red oaks. A similar-looking species, *L. cincinnatus*, grows on the roots of infected trees.

Joseph O'Brien, U.S. Forest Service

Laetiporus sulphureus

DO NOT eat any mushroom unless you are absolutely certain of its identity.

Laetiporus sulphureus

Laetiporus sulphureus

DO NOT eat any mushroom unless you are absolutely certain of its identity.

LOWLAND CONIFER ECOSYSTEM

Mike Ostry, U.S. Forest Service

Lowland conifer

Hollow Stem Larch Suillus   *Suillus cavipes*

Identification: Cap surface dark red-brown with dense hair; pore surface white-pale yellow with tubes radiating out from a hollow stem

Season of fruiting: Fall

Ecosystem function: Mycorrhizal with tamarack in bogs

Edibility: Choice

Fungal note: Squirrels often cache this species in trees (Fig. 2).

Suillus cavipes, top and bottom views. Note hollow stem.

Mike Ostry, U.S. Forest Service

DO NOT eat any mushroom unless you are absolutely certain of its identity.

Short-Stemmed Russula   *Russula brevipes*

Identification: Cap white-yellow, funnel-shaped; alternating long and short gills extending down the stalk

Season of fruiting: Summer-fall

Ecosystem function: Mycorrhizal with hardwoods, pine, and black spruce

Edibility: Edible, said to be choice if colonized by the orange fungus *Hypomyces lactifluorum* (bottom image)

Fungal note: Large groups of this mushroom can be overlooked because they are often partially covered by soil and leaf litter.

Joseph O'Brien, U.S. Forest Service

Russula brevipes

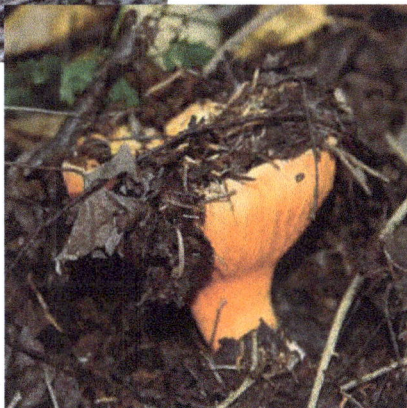

Mike Ostry, U.S. Forest Service

Russula brevipes parasitized by *Hypomyces lactifluorum.*

DO NOT eat any mushroom unless you are absolutely certain of its identity.

Swamp Death Angel   *Amanita brunnescens*

Identification: Cap light brown; veil and bulb present; white gills free from stalk

Season of fruiting: Summer-fall

Ecosystem function: Mycorrhizal with black spruce and tamarack in bogs

Edibility: Poisonous

Fungal note: *Amanita* mushrooms as a group are the most poisonous, accounting for almost all of the deaths caused by mushroom poisonings in the United States.

Mike Ostry, U.S. Forest Service

Amanita brunnescens

DO NOT eat any mushroom unless you are absolutely certain of its identity.

Larch Suillus   *Suillus grevillei*

Identification: Cap shiny, bright red-brown, smooth, sticky; lower surface yellow; prominent veil on stalk

Season of fruiting: Summer-fall

Ecosystem function: Mycorrhizal with upland tamarack

Edibility: Edible

Fungal note: An attractive, robust mushroom found only near tamarack.

Suillus grevillei

DO NOT eat any mushroom unless you are absolutely certain of its identity.

Identification: Cap dark brown, sticky; gills run down the thick stalk; lower surface white but turning black when spores are released

Season of fruiting: Summer-fall

Ecosystem function: Mycorrhizal with white spruce and other conifers

Edibility: Edible

Fungal note: Mushrooms in this group are also called slime caps.

Mike Ostry, U.S. Forest Service

Gomphidius glutinosus

DO NOT eat any mushroom unless you are absolutely certain of its identity.

Hedgehog Mushroom

Hydnum repandum (Dentinum repandum)

Identification: Cap buff-tan-dull orange with white-yellow teeth on the underside

Season of fruiting: Summer-fall

Ecosystem function: Mycorrhizal with conifers and hardwoods

Edibility: Edible

Fungal note: Spores are produced on the outside surface of the downward pointing teeth.

Hydnum repandum

DO NOT eat any mushroom unless you are absolutely certain of its identity.

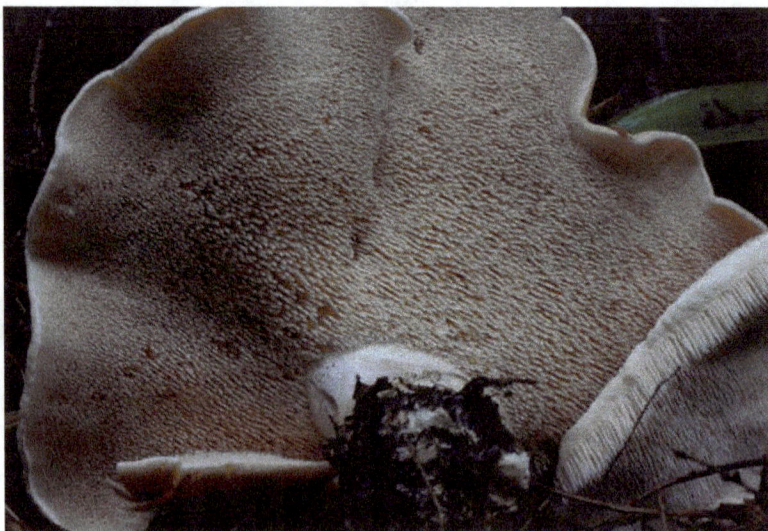

Hydnum repandum, bottom view.

Hydnum repandum, lower surface.

Milky Caps   *Lactarius volemus*

Identification: Cap rounded, center often depressed; all members of this group contain a latex that is exuded when the gills are cut

Season of fruiting: Summer-fall

Ecosystem function: Mycorrhizal with conifer and hardwood trees

Edibility: Not recommended; mushrooms with a latex that turns yellow or lilac color are poisonous

Fungal note: The edible *L. deliciosus*, found in conifer and mixed conifer-hardwood stands, has an orange cap that becomes stained green when bruised and contains a yellow-orange latex.

Lactarius volemus, top and bottom views. Note liquid latex on cut gill surface.

DO NOT eat any mushroom unless you are absolutely certain of its identity.

Identification: Cap smooth, bright red when fresh; evenly spaced white gills; stalk dull white and hollow

Season of fruiting: Summer-fall

Ecosystem function: Mycorrhizal with conifer trees

Edibility: Mildly poisonous

Fungal note: This species can be found in deep moss in bogs.

Russula emetica

Russula emetica, bottom view.

DO NOT eat any mushroom unless you are absolutely certain of its identity.

Yellow-Red Gill Polypore

Gloeophyllum sepiarium (Lenzites sepiaria)

Identification: Wooly, reddish brown shelf; yellowish-brown gills

Season of fruiting: Summer-fall

Ecosystem function: Brown cubical rot of conifers

Edibility: Inedible

Fungal note: This fungus can also decay coniferous wood products. *Gloeophyllum sepiarium* has both gills and pores and is thought to be a connecting link between the gill and pore fungi.

Joseph O'Brien, U.S. Forest Service

Gloeophyllum sepiarium

DO NOT eat any mushroom unless you are absolutely certain of its identity.

Hairy Cushion

Onnia tomentosa (Inonotus tomentosus, Polyporus tomentosus)

Identification: Brown to yellow, hairy-velvety, funnel-shaped or shelf-like on the ground at the base of trees

Season of fruiting: Summer-fall

Ecosystem function: Root and butt rot of pines and white and black spruce. A closely related fungus *Onnia circinatum* has been considered a variety of the hairy cushion. It causes a white pocket root and butt rot of pines and spruce. The species differ mainly by the shape of sterile bristle-like structures (setae) in their spore-producing areas.

Edibility: Inedible

Fungal note: Spread by root contact, this fungus causes stand openings.

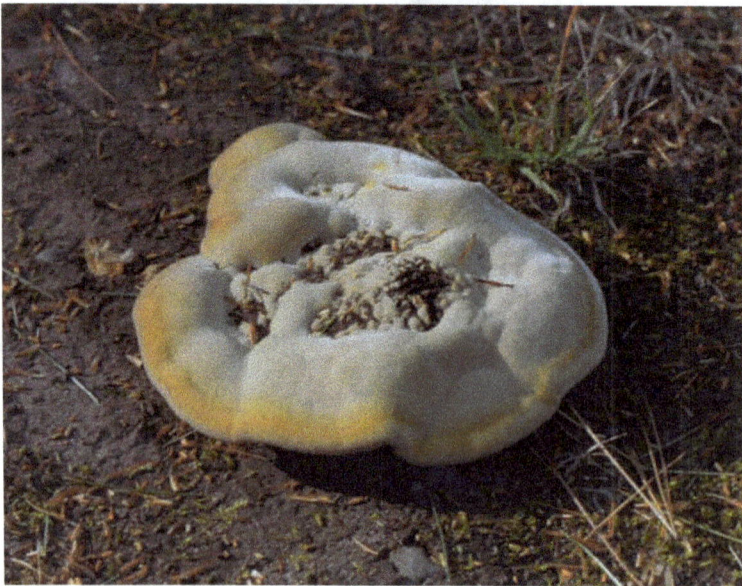

Joseph O'Brien, U.S. Forest Service

Onnia tomentosa

DO NOT eat any mushroom unless you are absolutely certain of its identity.

Coral Fungus   *Clavicorona pyxidata*

Identification: Multiple branched stalks, white-yellow, tips of branches forming a crown

Season of fruiting: Late spring-summer

Ecosystem function: Completes the breakdown of decayed wood

Edibility: Good when fresh

Fungal note: Spores form on the upright stalks of coral fungi and are thus unprotected from the elements compared to gill or pore fungi.

Mike Ostry, U.S. Forest Service

Clavicorona pyxidata

DO NOT eat any mushroom unless you are absolutely certain of its identity.

SUGGESTED REFERENCES

Arora, D. 1986. **Mushrooms demystified.** Berkeley, CA: Ten Speed Press. 959 p.

Bessette, A.E.; Bessette, A.R.; Fischer, D.W. 1997. **Mushrooms of northeastern North America.** Syracuse, NY: Syracuse University Press. 582 p.

Bessette, A.E.; Roody, W.C.; Bessette, A.R. 2000. **North American boletes.** Syracuse, NY: Syracuse University Press. 396 p.

Binion, D.E.; Stephenson, S.L.; Roody, W.C.; Burdsall, H.H., Jr.; Vasilyeva, L.N.; Miller, O.K., Jr. 2008. **Macrofungi associated with oaks of eastern North America.** Morgantown, WV: West Virginia University Press. 467 p.

Huffman, D.M.; Tiffany, L.H.; Knaphus, G.; Healy, R.A. 2008. **Mushrooms and other fungi of the midcontinental United States.** Iowa City, IA: University of Iowa Press. 370 p.

Lincoff, G.H. 1981. **The Audubon Society field guide to North American mushrooms.** New York: Alfred A. Knopf. 926 p.

McKnight, K.H.; McKnight, V.B. 1987. **A field guide to mushrooms of North America.** Boston, New York: Houghton Mifflin. 429 p.

Miller, O.K., Jr. 1981. **Mushrooms of North America.** New York: E.P. Dutton. 368 p.

Miller, O.K., Jr.; Miller, H.H. 2006. **North American mushrooms.** Gullford, CT, Helena, MT: FalconGuide. 583 p.

MYCOLOGICAL WEB SITES

http://www.mushroomexpert.com/

http://mushroomobserver.org/

http://botit.botany.wisc.edu/toms_fungi/

http://www.messiah.edu/Oakes/fungi_on_wood/index.htm

Ostry, Michael E.; Anderson, Neil A.; O'Brien, Joseph G. 2011. Revised
    February 2012. **Field guide to common macrofungi in eastern
    forests and their ecosystem functions.** Gen. Tech. Rep. NRS-79.
    Newtown Square, PA: U.S. Department of Agriculture, Forest Service,
    Northern Research Station. 82 p.

Macrofungi are distinguished from other fungi by their spore-bearing
fruit bodies (mushrooms, conks, brackets). These fungi are critical in
forests, causing disease, and wood and litter decay, recycling nutrients,
and forming symbiotic relationships with trees. This guide is intended
to assist in identifying macrofungi and provide a description of the
ecological functions of some of the most frequently encountered
macrofungi in aspen-birch, northern hardwood, lowland conifer, and
upland conifer forests in the Midwest and Northeast.

KEY WORDS: mushrooms, mycorrhizae, decomposers, pathogens,
conks, decay

www.ingramcontent.com/pod-product-compliance
Lightning Source LLC
Chambersburg PA
CBHW072209270326
41930CB00011B/2594